iPhone 12

SENIORS GUIDE

2023

The Ultimate Manual for the Non-Tech-Savvy to Master Your

iPhone 12 in No Time

John Halbert

Table of Contents

Introduction

Being in your sixties or seventies doesn't mean you can't use a new-generation phone. The iPhone is undoubtedly one of the most sophisticated mobile devices currently on the market. Not all seniors who already own a smartphone are in the market for a device with large buttons. You may wonder if you could purchase one for yourself and enjoy its popularly talked about features.

On the market, iPhones are currently among the most advanced mobile phones. Every year, Apple releases a new iPhone model, and it can be quite hard to distinguish between them. However, in this guide, we'll be focusing on the iPhone 12 because its feature addresses some important things you should think about when choosing the best iPhones for the elderly:

- **Accessibility:** Accessibility features include text magnification, voice dictation, and compatibility with hearing aids. However, they might also refer to a high-quality Retina display, a professional camera, or a wide variety of useful apps.
- **Cost:** It is well known that the average smartphone can cost well over a thousand dollars. When selecting the best iPhones for the elderly, we prioritize low-cost models.
- **Battery:** Battery life is a common complaint with devices like the iPhone. All the phones on this list have been proven to go a whole day without needing a charge during our tests.
- **Image quality:** The camera is one of the iPhone's most valuable features. Smartphones on this list make it simple to capture images of good enough quality to be shared online.

Because so many elderly people rely on hearing aids, we ensured that all of these phones were compatible. We didn't want them to deal with unwanted background noise or high-pitched feedback.

If money is no object, the Apple iPhone 12 is the best smartphone for seniors; however, the iPhone SE may be a better option for those on a tighter budget.

There is some good news, however: the iPhone isn't as complicated as it has been made out to be. You can use this phone until you're in your eighties without any problems. You can easily navigate the iPhone with the right information and set it up in ways that make it easier to use.

The iPhone 12 this year is the first to support 5G networks. It also has four distinct models and four sizes available. Apple's iPhone 12 is the company's fastest, thinnest, and first 5G-capable smartphone. It has a faster processor, a better camera, and the magnetic MagSafe charging method. When choosing between the four models, screen size is the deciding factor. The 12 mini, with a 5.4-inch display, is the smallest iPhone. This is the best option if larger phones make you uncomfortable or have trouble keeping them in your hand. Most people will be satisfied with the iPhone 12. The iPhone 12 Pro is also an excellent option if you want to upgrade your camera quality. At the same time, the Pro Max is a fantastic choice for grandparents because of its large 6.7-inch display and surprisingly high-quality camera.

As mentioned earlier, the iPhone 12 is compatible with the most recent 5G networks, but 5G coverage is patchy across the United States, making the upgrade to the iPhone 12 unappealing. It could take at least a year for 5G to become widely available.

This is the best iPhone available if money is no object, regardless of age. With the cheapest iPhone 12 mini costing $699 and the most expensive iPhone 12 Pro Max costing an eye-watering $1399, the price tag may be prohibitively expensive for some customers.

Chapter One: Purchasing Your iPhone 12

Purchasing the iPhone 12 is not a difficult process. You can purchase your phone from various online retailers and other outlets. It is, however, strongly advised to purchase your phone from reputable retailers such as Apple or licensed resellers. The following will teach you how to choose an iPhone 12.

Factors To Consider When Purchasing Your iPhone 12

1. Don't automatically go for the newest phone. If you're on a tight budget, it might not be wise to acquire the most recent iPhone model in shopping mode. Apple iPhones from a few years ago still capture great photos and perform well.

2. You can purchase an iPhone with less storage space and save some money. A smartphone's storage can be increased with a computer or laptop.

3. Online stores provide substantial savings on purchasing an iPhone when you trade in an older model. Some vendors may provide rebates on top of the money you could get from selling your used machinery to encourage a purchase.

4. Invest in refurbished or used cell phones. You might save money by purchasing a gently used iPhone 12 or 13 from shopping mode. Some reliable sellers provide used shopping mode iPhone models with a vendor warranty.

Questions To Ask When Purchasing Your iPhone 12

1. What are the different models of the iPhone 12?

There are two main versions of the iPhone 12: the base model with 128GB storage capacity and 6 GB RAM and the 256GB version with 8 GB RAM. Both variants come in Silver, Space Grey, Gold, Rose Gold, Jet Black, and Midnight Blue.

2. How much does the iPhone 12 cost?

You can get the iPhone 12 for $699 (128GB) and $849 (256GB). On the apple store, the iPhone 12 Pro goes for $999, and the iPhone 12 Pro Max goes for $1099

3. Where can I buy the iPhone 12?

Apple sells the iPhone 12 directly through its website at www.apple.com/iphone. It also offers the device through third-party retailers like Amazon, Best Buy, B&H Photo Video, Newegg, Target, Walmart, and others.

4. Can I use my existing contract when buying the iPhone 12?

Yes, you can use your current carrier's plan to buy the iPhone 12.

5. Is there any difference between buying the iPhone 12 on contract versus paying full price?

Purchasing the iPhone 12 on contract allows you to pay monthly installments over 24 months. Until the entire amount is paid, you'll have to make monthly payments. You will not be required to make monthly payments if you pay the full retail price. Instead, you'll have to pay a single lump sum.

6. How do I know which variant of the iPhone 12 I should buy?

Consider what characteristics are most important to you before choosing the iPhone 12 (128GB or 256GB). For instance, if you want a phone with a long battery life, the 256GB version is preferable to the 128GB one. However, the 128GB model is best if you're more concerned with having enough room for apps and games.

7. Do I need to worry about getting stuck with a bad unit?

No, you don't need to worry about getting a defective unit. All units sold by Apple are thoroughly tested before they leave the factory.

8. How long does it take to get the iPhone 12?

It usually takes 2 - 3 weeks for delivery after placing your order.

9. Does the iPhone 12 support wireless charging?

Yes, the iPhone 12 supports Qi wireless charging.

10. Does the iPhone 12 support fast charging?

Yes, it supports both standards and reverses wireless charging.

11. How many cameras does the iPhone 12 have?

The iPhone 12 has three rear cameras: A wide-angle camera, a telephoto lens, and a depth sensor. There is no dual-camera setup.

12. Which color option is best for me?

If you buy the iPhone 12 in silver, go with the matte finish. Matte finishes are excellent at resisting fingerprints and scratches. They also look good in low-light situations. You should get the glossy finish if you want to buy the iPhone 12 in gold. Glossy finishes reflect light and make surfaces appear brighter. They may, however, be susceptible to scratches and fingerprints.

Chapter Two: Setting Up Your iPhone

You just got your new iPhone 12 and have no idea how to use it. First, it is critical to recognize that the iPhone 12 is not difficult to set up. It is an amazingly simple procedure. After unboxing the new phone, remove the phone and charger and ensure the phone is fully charged before proceeding with the operation.

You should also look for the instruction booklet, which includes a detailed guide. When inserting your Sim Card into the sim tray on the side of the iPhone, you can use the sim ejection tool, a small piece of iron that looks like a paper clip and is included in the box.

You need an internet connection when setting up your new iPhone 12. Therefore, you need to connect it to Wi-Fi or a cellular data network. If you want to use a cellular data network via your local network provider, locate the sim tray at the left side of the iPhone and use the sim ejection tool to poke the hole for the sim tray to pop out. Place the sim as fit, then insert it back into the iPhone.

How To Turn Your Phone On/Off

Starting with the iPhone 12, switching it on and off can be a chore. It's simple to turn on an iPhone, but it's more difficult to turn off because extra safeguards are in place to prevent accidental deletion. This guide will show you three different ways to turn off and on your iPhone 12. The iPhone 12 has three distinct on/off modes:

1. **How to use the side button to turn on the iPhone 12.**

You just purchased an iPhone 12 and are unsure how to activate it. Like on other devices, the Apple logo can be displayed by simply holding the Side Button on the right edge of your iPhone. While you wait for the iPhone to boot up, let go of the button.

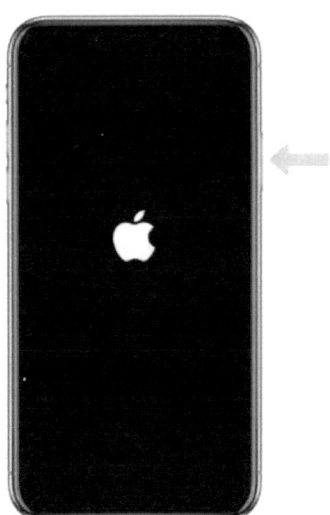

You will be asked to enter your iPhone password and SIM PIN. If you recently purchased an iPhone, you should customize it by changing the language displayed on the screen or adding a unique ringtone.

2. **Your iPhone will turn on when you plug it in.**

Even if your iPhone 12's side button is broken, there is a way to turn it on. Simply connect the charging cable to the wall, a portable power source, or a USB port on your computer, and then connect your phone. The iPhone should now be activated. When you plug in your iPhone to let it charge, it automatically turns on.

Remember that if your iPhone's battery is at or below 3%, it will not turn on until fully charged. The same is true for the next method.

3. You can turn on your iPhone using a wireless charging pad.

The most recent iPhones can be charged wirelessly. If you place your iPhone on a wireless charging pad, it will turn on automatically if the power button fails or if you simply want to use it without touching any buttons. When you place your phone on the pad, it will begin to charge.

How To Turn Off Your iPhone

A few ways exist to turn off your iPhone if it is not responding, moving slowly, or needs to conserve power because you are far from a charger. We should start with the most well-known. Previously, you had to tap and hold the power button to turn off an iPhone. Apple changed its policy after discovering that users could unintentionally turn off their devices while carrying them.

The iPhone 12 can be turned off by pressing the Power button and, on the left side, either the Volume Up or Down button. An emergency menu will appear if you continue to hold this combination for a few seconds. Don't let go just yet! The iPhone takes a picture when the Side and Volume Up buttons are pressed simultaneously, so keep holding the buttons to ensure the Slide to Power Off screen shows up.

1. Tap and press the Power button and Volume Up/Down buttons.

You can release both buttons as soon as the emergency screen appears. A "slide to power off" button is at the screen's top. To turn it off, swipe your iPhone on the on/off switch. To complete the action, slide the button to the right.

Be cautious when employing this strategy because your emergency screen alerts others that you are in trouble. Even though the next two options are riskier, they may be better for you if this is the case.

2. Using the power-off options in the system's menu.

You can also alternatively turn off your iPhone from the Settings menu if you don't like the tap-and-hold method or if the buttons don't work. Navigate to the Settings menu > General > Shut Down at the bottom of the screen. Slide your iPhone back to the "power off" position to confirm that you want to turn it off. Your iPhone will turn off after a short time.

WARNING: If you cancel any of the above methods of turning off your iPhone, it will lock down, and you will be unable to use Face ID until you enter your passcode. This feature may save your life in a dangerous situation, but it can be annoying if you accidentally activate it. This is simple if you're coming from Android, where similar shortcuts exist.

Chapter Three: Setting Up Your iPhone 12

After you've unboxed and turned on your new iPhone 12, the next step is to configure it. While you should know a few things before beginning setup, we'll review everything in this article. When you see the word "Hello," most likely in another language, swipe up and proceed as follows:

Step 1: Set Up Language and Country.

After turning on your phone, you should configure your language and country settings. This allows you to use Siri and other iOS features without translating them into English or your native language. When the language options appear, select English, and the iPhone will automatically select your country. Similarly, options would be presented; simply select your home country.

Step 2: Review the Quick Start Prompt.

Quick Start is a feature that allows you to transfer content and data to a new device from an old one while the new one is being set up. On your old device, tap Continue. Hold your new iPhone up to your old device's camera. Enter the passcode from your old device into your new device.

Step 3: Connect to Wi-Fi.

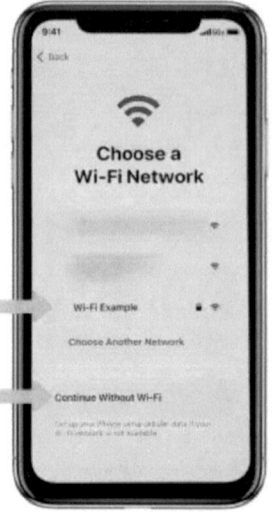

When you turn on your new iPhone 12, you must first connect to Wi-Fi. This will enable you to connect to Wi-Fi networks at home and work and any public hotspots you may encounter. You can also use Wi-Fi Assist to switch from cellular data to Wi-Fi if your battery runs out. Tap the Wi-Fi network you want to connect to, then enter your password. Click the "Create Password" button to generate a password if you don't have one.

Step 4: Look over the "Data & Privacy" message, and then click "Continue" or "Learn More."

This is where you review critical information about how your data is used. Understanding what happens to your data is critical because it may impact your privacy. Apple, for example, collects location data to improve Maps but does not share it with anyone else unless you give permission.

If you're okay with sharing your location data, proceed to the next step.

Step 5: Set up Face ID now or click Set Up Later to skip this step.

The facial recognition system built into the iPhone 12 is Face ID. It's designed to make unlocking your iPhone faster and more secure than ever. For things like Face ID, you'll need a passcode. A prompt would come up that requires you to enter a passcode. Simply enter the Passcode you want to use, then enter it again to confirm. You can skip this step later by clicking "Set Up Later."

Step 6: Create Your iCloud Account.

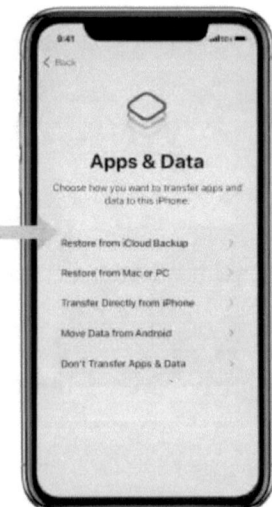

When you boot up your iPhone 12, you'll be prompted to sign in to iCloud. Signing into iCloud gives you access to your apps, music, photos, documents, and other files across multiple devices.

Tap the "Sign In" button to create an account. Enter your email address and password, then press the Next button.

Step 7: Choose What Apps Can Access Your Data.

The next screen asks you to choose which apps can access your data. Tap on each app to see what kind of data they can access. For example, if you want to keep your contacts private, you'd tap on Contacts. For everything else, tap on OK.

Step 8: Add Family Members' Information.

After you've decided which apps have access to your data, you'll be asked if you want to add family members to your existing iCloud account.

If you already have an iCloud account, tap "Add Family Member" and follow the instructions to add your family member. If you don't already have an iCloud account, click "Get Started Now" to set one up.

Step 9: Review Your Settings.

Investigate your new iPhone 12's settings. This section explains what each option does:

- Read the Terms and Conditions and then click "Agree."
- Examine the message "Keep Your iPhone Up to Date," then press the "Continue" button.
- Check out the Location Services prompt and select the one you want.
- Review the Siri prompt, then click Continue to proceed with the setup instructions or Set Up Later in Settings.
- Check the Screen Time prompt, then click Continue, follow the on-screen instructions, or go to Settings and click Set Up Later.
- Examine the iPhone Analytics message, then click Continue.
- Look over the App Analytics prompt and select the desired option.

Step 10: Select the desired appearance, then select Continue.

Select the color scheme you want for your iPhone 12. You can also change the wallpaper from here. Choose between two different wallpapers. Tap on the slider to adjust the volume.

Step 11: Select the desired display zoom, then select Continue.

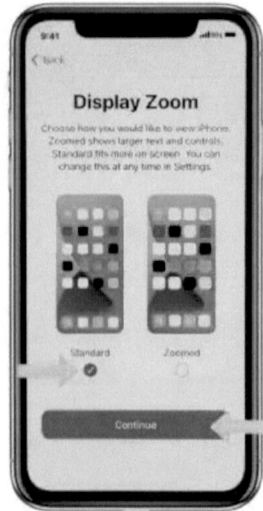

You should choose the zoomed appearance option because it allows you to see letters and use the iPhone 12 features more effectively as you age. When you've finished all of these steps, click continue. You'd see the word "Welcome" on your screen and would swipe up to begin using your device.

Charge Your Phone At Home or Using PC Cable

The iPhone 12, iPhone 12 Pro, and iPhone 12 mini no longer come with Apple's conventional USB charging brick. Only a USB-C to Lightning cable may be found in the packaging. You simply need to buy a charging brick or use your wall adapter from your previous iPhone to charge your iPhone 12.

How to Charge Your Phone Using a Computer

You may have heard you can charge your mobile device using a computer, but how exactly do you do it? The process is simple and involves plugging in a USB-C to USB-A adapter. The adapter allows you to connect your iPhone to a computer via a standard USB port. Once connected, download and install the right software on your PC to begin charging your phone.

How to Lock/Unlock the Screen

Pushing the power button on the top of your iPhone will prevent unauthorized access. The lock screen will remain locked until the correct passcode is entered. When you're not using your iPhone, it locks for security, enters a power-saving sleep mode, and turns off the display to save battery life. There are three methods for unlocking your phone:

1. Using Face ID to unlock your iPhone.

If you own an iPhone XS or XS Max and forgot to activate Face ID during setup, you can do it by going to Set up Face ID. Lifting the iPhone to your ear or tapping the screen will wake it, so just glance at it. The lock icon changes from closed to open when an iPhone is unlocked.

A glance at your iPhone will tell you whether you need to tap the screen or lift it to your ear to wake it up. The lock icon changes from closed to open when an iPhone is unlocked. Swiping up from the bottom of the screen will bring you to this feature.

To lock the iPhone, press the side button once more. It locks if you don't use your iPhone for a minute or more. If you enable Attention Aware Features in Settings > Face ID & Passcode, your iPhone will not lock or dim when it detects that you are looking at it.

2. Using Apple's iPhone Touch ID for unlocking.

Turn on Touch ID after buying an iPhone if you missed the chance during Setup. To unlock iPhone, press the Home button with the finger you set up for Touch ID. Tap the side button once more to lock your iPhone. An iPhone locks itself after about a minute of inactivity.

3. Enter the passcode to unlock iPhone

Set a passcode on your iPhone if you did not do so during the initial setup. Swipe up from the bottom (if using Face ID) or tap the Home button to get to the Home screen from the Lock Screen (on other iPhone models). Now enter your secret code. Tap the side button once more to lock your iPhone. An iPhone locks itself after about a minute of inactivity.

Making Emergency Calls

In the event of a crisis, you can quickly alert significant individuals and contact emergency services using your iPhone (provided that cellular service is active), thanks to its built-in SOS feature is available.

How It Works

Your iPhone automatically dials the local emergency service when you use SOS to place a call. You should select the required service in some nations and areas. For instance, you can select fire, police, or ambulance on the Chinese mainland.

Additionally, you can add emergency contacts. Unless you decide to cancel, your iPhone sends a text message to your emergency contacts once an emergency call finishes. Your emergency contacts will receive an update when your location changes after you enter SOS mode on your iPhone, which communicates your current position.

When SOS shows in the status bar of your iPhone, a cellular network is accessible for making emergency calls. Find out more about the symbols and status icons on your iPhone.

How To Make an Emergency Call on iPhone 12

Tap and hold the side button and one of the volume buttons until the Emergency SOS slider appears.

Move the Emergency Call slider to contact emergency services. If you continue to hold the side and volume buttons, a countdown begins, and an alert plays instead of moving the slider. If you continue to hold the buttons after the countdown, your iPhone will dial 911.

End an Emergency Call

If you start the countdown by accident, you can stop it. On an iPhone 12, release the volume and side buttons and tap Stop Calling.

Include Emergency Contacts

1. Tap your profile image in the Health app after starting it. Toggle Medical ID.
2. Scroll to Emergency Contacts after tapping Edit.
3. To add a contact for emergencies, tap the Add button. Tap a contact, then include their connection.
4. To save your changes, tap Done.
5. Emergency services cannot be designated as an SOS contact.
6. In the Health app, create a Medical ID.

How to Remove a Contact from Your Emergency List

1. Tap your profile image in the Health app after starting it. Toggle Medical ID.
2. Scroll to Emergency Contacts after tapping Edit.
3. When a contact is selected, touch Delete after tapping the Delete button.
4. To save your changes, tap Done.
5. Modify how you call

You can call emergency services immediately if you have an iPhone 12 model. When you try to make an emergency call while Call with Hold or Call with Five Taps is enabled, your iPhone starts a countdown and plays an alarm. When the countdown expires, your iPhone automatically contacts emergency services. To activate these settings:

1. Open your iPhone's Settings app.
2. Select Emergency SOS.
3. Call with five taps or hold, then dial.

Note: You can still call using the Emergency SOS slider even if Call with Hold or Call with five taps are disabled.

How to Switch to Silent Mode

Adjusting the volume, setting a new ringtone, and altering the sound for notifications are straightforward processes.

1. Assessing the sound settings.

Start the Settings app from the home screen, then go to the Sound & Haptics tab.

2. You may need to turn up the volume.

Tap the volume keys on the left side to adjust the volume of the phone's audio output or an incoming call. The volume can also be adjusted from the Sounds & Haptics menu. Simply use the Volume slider and move it to the desired level. To enable or disable volume control via buttons, toggle the Change with Buttons switches.

3. Modify your phone's settings to vibrate and silent modes.

On your iPhone, slide the Ring/Silent switch to the left to turn on quiet mode. On the Sounds & Haptics screen, the Play Haptics in Ring Mode and Play Haptics in Silent Mode switches to enable and disable haptics, respectively.

4. Edit the alert tone.

Find the alert you need to customize the sound on the Sounds & Haptics screen. When you're done, click the back button.

5. Put your device into DO Not Disturb mode.

Scroll down from the upper right corner of the home screen to open the Control Center to easily enable Do Not Disturb. Select "Focus" and "Do Not Disturb" from the menus. When Do Not Disturb is enabled, all messages, calls, and notifications are silenced. To learn more, contact Apple's support team.

Users of the iPhone 12 can enable Do Not Disturb. When the feature is enabled, a Do Not Disturb icon appears on the Lock screen, notification bar, and Control Center.

Setting Up Your Home Screen

Apps and widgets on the iPhone's Home Screen can be moved to other Home Screen pages, restored to their original layout, and more. This is especially important for the elderly, so only relevant apps are displayed to them, especially on the home screen.

How to Move Widgets and Apps on the Home Screen

You can rearrange your home screen icons and widgets on iOS 13. Here you can choose whether to move apps or widgets and how many rows and columns you want each app or widget to span. Then tap the + button to add a row or column. Tap and hold to resize rows and columns. Dragging an icon off the grid deletes it permanently.

How to Change Background and Lock Screen Photos

Do you want to customize the background on your iPhone? In iPhone 12, it's easier than ever to add a new background to your device. You can use Apple's default background or upload your image. Here's how to go about it:

How to Change Wallpaper

Launch the Settings app, go to General, then Wallpaper, and finally, Add New Wallpaper. You can choose a wallpaper from one of four categories—People, Places, Animals, or Photo Shuffles—and even further customize it.

If you want to use something other than a standard image, you can search the Apple App Store for a free app that allows you to take photos and create collages; once you've decided on one, download and install it on your iPhone. Launch the app, tap the plus sign next to the camera icon, followed by Choose Photo. Then, tap Browse All Photos and scroll down to find a photo you like. Finally, press the Select button.

Once you're done picking out a photo, tap Save and Done. Your new wallpaper will now show up on your home screen.

How to Change Your Phone's Background Picture from the Lock Screen

You can now change your Lock screen wallpaper directly from the Settings app. This feature works best with images taken with Apple's True Depth camera system. You can use the same technique to set any image as your Lock screen background. To do this, follow these steps:

1. Open the Settings app.
2. Tap Wallpaper & Brightness.
3. Tap Choose Your Background Image.
4. Touch and drag the photo you want to use onto the preview area.
5. When you find one you like, tap Set As Wallpaper.

Customizing Your Lock Screen

Apple introduced several features in iPhone 12 that allow users to customize the appearance of their lock screen. Filters and widgets are now available, as well as the ability to change the colors of your current wallpaper.

To access the wallpaper settings, launch the Settings app and select Wallpaper. Tap the Customize Current Wallpaper section from there. In this section, you can select a background image from your device's camera roll, iCloud photos library, or Photos library. You can then add a filter, widget, or style.

The same feature can also be used to add new wallpaper. Simply tap the plus sign next to the word Choose in the Preview box to select a photo from your photo library. Then, to save it, tap Add Current Wallpaper.

How to Create App Folders

Did you know that you can organize the Home Screen pages of your apps by creating folders? You can accomplish this by following the steps below:

Hold down the Home Screen wallpaper until the icons begin to shake. Hold down one of the icons until it begins to wobble. Then, tap and hold the icon you want to move again. When this happens, a menu will appear, and you can either delete the original app or drag it to a different location.

How to Move an App from a Folder to the Home Screen

You can easily move an app from a Folder to the Home Screen. This makes it easy to find and launch the app. To do this, follow these steps:

1. Open the App Drawer.
2. Tap the folder icon next to the app you want to move.
3. Touch and drag the app to the Home Screen on the screen that appears. When you release the app, it moves there.
4. Then, add the folder where you want the app to show up.

5. Now go to the Home Screen and swipe down to view all apps on your phone. Find the one you just added and tap it to open it.

How to Setup an Existing Account or Create a New One with Apple iCloud

Apple has introduced "iCloud," a cloud-based storage system for all the data and files you store on your device. The main benefit of using this service is that you can access all of your data from any other iOS device or Mac computer. If you already have an Apple ID on one of these devices, you can sign in to iCloud.

Setting up an Apple iCloud account is a simple process. You must enter basic information about yourself, such as your name, email address, phone number, birth date, and password. After you have completed all of the required fields, you will be prompted to enter security questions to confirm your identity.

Once you have completed the setup process, you will receive a confirmation code via text message. This code must be entered into the settings app before logging into iCloud.

Steps to Set Up Apple iCloud Account on iPhone

1. Open the Settings app.
2. Tap on iCloud.
3. Enter your Apple ID and Password.
4. Verify your Identity using Security Questions.
5. Tap on Sign In.
6. Enter the Confirmation Code sent to your mobile number.
7. Log in successfully.

How to Set Up the App Store

The app store can be accessed by touching the App Store icon on the iPhone's Home screen. You may access additional details about an iPhone app by tapping on it. To set up the app store, you can take the following steps:

1. Tap the app store icon.
2. Tap on the "Add profile" button.
3. Enter the name of the new profile (e.g., Jackson).
4. Select the "Create New Apple ID" option.
5. Enter the email address associated with your Apple account.
6. Click on the "Next" button.
7. Enter the password associated with your Apple account on the next screen.
8. Confirm the details by clicking on the "Sign In" button.
9. Now tap on the "Continue" button. Once the process completes successfully, you will be redirected back to the home page of the App Store.

How to Set Up Your Credit Card

Not every app and product on the app store is free; if you use these Apple products, you must purchase them. Adding a credit card to your iPhone 12 allows you to purchase any app from the App Store.

To add a credit card to your iPhone, go to Settings and select Wallet & Apple Pay. This can also be done by tapping the "Wallet" icon at the bottom of the screen. After you've opened the wallet, scroll down to the Credit Card section. Tap on that, then follow the steps below:

1. Add Your Credit Card

The first step is to enter your credit card information. To do so, launch the Wallet app and select the Credit Card option. A new window will appear containing all the information required to add your credit card. Find your credit card number by scrolling through the list. If it doesn't appear, enter the last four digits of your credit card number. Your full name or other identifying information may be required. When you're finished, click the Save button.

2. Verify Your Information

You should verify your credit card information after you have entered it. This means you must enter your credit card information. You would also need to enter your credit card's security code. Enter the code, then press the Verify Code button. You can now exit the Wallet app.

3. Set Up Automatic Payments

You can now set up an automatic payment if you don't already have one. Return to the Wallet app and select the Payment Method option. You can choose whether to pay each month automatically or whenever you buy something from this page. If you choose the monthly option, you must enter the monthly payment amount for your credit card.

For example, if you wanted to put $50 on your credit card each month, you would enter 50 in the Amount box. Then you'd press the Next button. If you selected the weekly option, you would simply enter how much you want to spend each week. To do this, you would click on Weekly Spending and enter the amount you want each day. Clicking on the Next button will take you to the next page, where you can save your settings.

4. Use Your Credit Card

You can use your credit card now that you've set up automatic payments. Reopen the Wallet app and select the Credit Cards option. You will see a list of all the cards you have saved. Choose your credit card, then click the Make Purchases button. This will display a confirmation screen with the items you just purchased.

When making online purchases, you can also use your credit card. Launch the Wallet app once more. Select your card by tapping the Credit Cards button. Then, press the Buy button. You'll see another confirmation screen that shows you what you just purchased.

If you ever forget your credit card, you can always call customer service to get help. They will ask you to enter your security code, and they will cancel your card. Afterward, you can re-add your card using the steps above.

Downloading an App on the App Store on iPhone 12

Downloading apps from the app store is pretty straightforward. Following these steps, you can download any application you want on your iPhone 12.

1. Tap on the 'Search' tab at the top right corner of the screen.
2. Type the name of the app or game that you want to install on your device.
3. Tap on the search icon located below the text box. The list of results will appear on the screen.
4. Scroll down until you find the app or game that interests you.
5. Tap on the "Install" button. A pop-up window will appear, asking whether you want to continue.
6. Tap on the "Yes" button to proceed further. Your iPhone will now start downloading the app.
7. When the installation process is complete, you will see a message saying the app has been installed successfully.

How to Update an App on iPhone 12

It is simple to update your apps on an iPhone 12. For older people, auto-update is the best option; this way, you don't have to manually update your applications whenever a new update is available. There is no need to download the updated version from the App Store if you are running iOS 13 or later. All updates will be downloaded directly to your device without the need to visit the App Store.

How to Uninstall an App on iPhone 12

To remove apps from an iPhone 12, go to the home screen. Follow these steps instead:

1. Locate the program you want to uninstall on your iPhone's main screen.
2. Tap and hold the icon to access the app's settings. You can keep pushing after the menu appears to access more options. Your apps will reorganize, and the menu will be removed. You can now uninstall the app by tapping on it.
3. Simply tap and hold the app's icon to access its settings. Tapping again once the menu appears is another option. The menu vanishes as soon as you do this, and your programs rearrange themselves. If this occurs, choose the option to uninstall the program.
4. A prompt will appear if you want to uninstall the program or transfer it to another device. Before you can uninstall an app that uses iCloud to store data, a prompt will appear asking if you want to delete the data along with the program. You can access the data after reinstalling the app if you do not delete the data. If the unwanted program's data is saved in iCloud,
5. When you uninstall an app, it goes away. Do this for each unwanted program you want to remove.

Chapter Four: Communicating Using Your iPhone 12

Send and Receive Messages on iPhone 12

iMessage allows Apple users to communicate with one another via Wi-Fi or cellular data. Blue balloons represent conversations on iMessage. You can also send and receive messages via SMS or MMS from your mobile carrier.

You can send photos, videos, and other media types in an iMessage. Sending a read receipt notifies the sender that you have seen their message and can see when they are composing it. All previously sent messages can be retrieved and modified. For added security, iMessages are encrypted before being sent.

Furthermore, there are blue bubbles for multimedia messages and green bubbles for text messages. SMS messages only have text, while MMS messages can only have pictures, sounds, and videos.

How to Send and Receive messages

You can now send text messages to multiple recipients at the same time. Open Messages, tap the person icon next to the text box, followed by Send Message. If you have Dual SIM enabled on a compatible model and want to send an MMS message from a different number, tap the line shown and select a different number. Then enter the recipients' phone numbers, contact names, or Apple IDs.

A new conversation begins when someone contacts you for the first time via a messaging or chatting service. Along with the message, a "Conversation" button will appear. By tapping this button, you can start a new chat.

The messages will be added to the conversation thread you already have in Messages with that user. Now it's a lot easier to follow the conversation along.

Select the desired chat from the Conversations list and touch the Join button. A blue dot marks your previous progress.

Finding Friends and Family on iMessage

To find people and content in conversations, open the search bar above the Conversations list. Searching brings up suggestions like contacts, links, photos, and more. You can now respond to a specific message in a group or individual conversations. Touch/hold a message in a conversation, then tap "reply." Then write your response and tap "reply." If someone else replies to your message, you'll see it appear in blue next to your original message.

If you double-tap the message bubble, you can choose one of several different exceptions to send your response. For example, you could say "thumbs up," "heart," or "thanks." To ensure everyone sees your reply, you can use the @ symbol to mention people in the conversation.

Managing Contacts

The iPhone has a built-in address book where you can track all your loved ones' phone numbers. Thanks to this, you may get in touch with people fast via phone calls, emails, texts, and other forms of electronic communication without having to dig up their contact details again manually.

Adding, Saving, Deleting, and Importing Contacts

You can add contacts manually or automatically using information from your address book. To manually add a contact, tap the + icon next to the name of a friend or family member. Enter your full name, phone number, email address, and relationship to you when prompted. To save the contact, tap Done.

Tap Add Contacts to automatically add a contact. You can import contacts from your previous device or iCloud. Select Import All Contacts if you're importing contacts from a previous device.

Turn off Automatic Import if you don't want anyone else to import your contacts. Then, tap Clear All to remove all your contacts. Tap Edit to manage your contacts. You can change contact information, such as names, addresses, and phone numbers.

When adding contacts, you can specify how frequently you want to be notified of changes. For example, if you recently added a contact, you should receive daily notifications. iOS sends you notifications once per week by default.

Making Video Calls on iPhone 12

With an internet connection and your new iOS device, you can make and receive calls using Apple's FaceTime video calling app. This necessitates having a valid Apple ID and pre-configuring FaceTime. After setting these up, it's simple to begin making and receiving calls.

Open the FaceTime app and tap the + button in the lower left corner to make a call.

Select Call Someone Else. In the space provided, enter the name or phone number of the person you want to speak with. Then press the Start Calling button. A green check mark appears beside the individual's name. To end the call, select End Conversation.

How to Make a FaceTime Call

You can either call someone or start a group chat. If you want to call someone, enter their phone number in the entry box at the top of the screen and press Call. Alternatively, near the top of the screen, tap New FaceTime and enter their name or email address. Make a video or audio call by tapping Make Video Call or Audio Call.

But what if you don't have someone's phone number? Or are you attempting to contact someone who does not own an iPhone? These are common issues that elderly people encounter when using the iPhone 12.

Apple's answer is "Siri." With Siri, all you need to do is speak to your device. You can use phrases like "FaceTime Jude" or "Call Daniel." In the upcoming chapters, we will convert how to set up and harness the Siri feature to make the best out of your iPhone 12.

Receiving a FaceTime Call

When you receive a FaceTime call, there are four options: accept, decline, set a reminder to call back, or send a text message. You'll never want to decline a FaceTime call because it interrupts whatever else you do. Plus, you might miss out on a great opportunity to make a sale.

Rather than accepting or denying a FaceTime call, the End & Accept button will appear if you are already in the middle of a call when one comes in. This lets you end the current call and connect to the incoming FaceTime call without picking one over the other.

Start a Video Call from a Message Conversation

In a Messages conversation, you don't have to tap "FaceTime." You can start a FaceTime call directly from within the chat window. To do this, tap at the top right of a Messages conversation. Then, tap either FaceTime Audio or FaceTime Video. This opens up a list of options for choosing the type of call you want. If you want to initiate a video call, tap FaceTime Video. If you want to send audio only, tap FaceTime Audio.

If you've got someone in your contact list who isn't already on your iOS device, you'll see a prompt asking whether you'd like to add them. Once you've added the person, you can begin making calls.

You can also use FaceTime Audio to call anyone else on your iPhone or iPad. Just tap the name of the person you want to call, and you'll be able to make a voice call.

Sending Emails

You must use the regular mail app to send emails from your iPhone 12. It's simple: write, send, receive, and schedule emails from your other accounts.

Creating an Email Message

To begin, tap the + icon or New Mail. After that, type the reason for the email in the heading box and press enter.

The next step is to add recipients. Tap the To field and enter the name(s) of each person to whom you want to send a message. Then select Add Recipients.

Mail suggests contacts and addresses as you type for people who have multiple email addresses. If the contact you're looking for isn't listed, try tapping the Contact button at the bottom of the screen.

Lastly, you want to compose your message. Tap the input box and type your message.

You can change the formatting, choose from different fonts and colors, make lists easier to view, and more. Once you are done typing, tap send, and your email will be delivered to your designated recipients.

How to Fill Predictive Text/Deactivate/Personalize

As you type on the iOS keyboard, predictions about what words might come next appear. You'll also see suggestions based on your recent activities and app data. For example, if you recently searched for "laundry detergent," one of the results will be laundry detergent. You'll see Good Morning if you just typed "good morning." And if you've been typing messages to friends, you'll notice their names appear. This feature can help you avoid typos, save time, and communicate more effectively with others.

How to Accept or Reject a Suggestion

When you type text into a phone keyboard, you can choose whether to accept or reject a suggested word or emoji. You can also select one of three ways to handle a highlighted prediction:

1. Tap the word to insert it.
2. Enter a space or punctuation mark to insert it.
3. If you don't want to use the suggestion, you can delete it by tapping the X icon.

How to Turn Off Suggested Predictions

1. A prompt showing the keyboard setting will appear shortly when you edit text, touch, and hold a suggestion for a few seconds.
2. Tap keyboard settings, then turn off predictive.
3. iPhone may still suggest corrections for a misspelled word.
4. Type a space or punctuation mark or tap return to accept a correction.
5. To reject suggestions, tap x. If you reject the same correction several times, iPhone stops suggesting the correction.

WhatsApp on iPhone12

WhatsApp has quickly grown to become one of the most popular messaging apps. With over 1 billion active monthly users, you're probably already familiar with it. It is one of the best ways to stay in touch with family and friends because you can see what they are up to on their status and interact with them right away.

Installing WhatsApp

Getting WhatsApp on your iPhone 12 isn't difficult; you can follow these steps to install the app on your IOS device:

1. Open the App Store app on your iPhone 12 Pro and select "Search."
2. Type "WhatsApp" and hit enter. You will see a list of applications related to WhatsApp. Select one of them and tap on "Install."

3. Wait for the installation process to complete. Once completed, launch WhatsApp on your device.

If you want to use WhatsApp Business to communicate with your customers, the application is available in the Apple Store. You can get it just like any other WhatsApp application. Once installed, search for "WhatsApp Business" to access the WhatsApp Business interface, where you can create new groups, invite people to join them, and begin communicating with your customers.

Both apps can be installed on the same device. Please contact the WhatsApp team if you have questions about using WhatsApp Business. Furthermore, the WhatsApp team provides training courses to help businesses worldwide understand the app's features.

Using WhatsApp

The basics are simple: once you download the app, you'll see three icons on your home screen: Messages, Chats, and Settings.

Messages are where your text, call, and video chat with others. Chats let you start conversations with multiple contacts at once. And Settings provides access to account settings and data usage information.

To make a phone call, tap the Chat icon, select the person you'd like to talk to, and hit Call.

To send a photo or video, tap the Chat icon again, select the type of media you'd like to add, and hit Send.

You can also record audio and video directly within the app. Just open up the camera and tap Record Video or Start Audio Recording. When you're done recording, tap Stop Recording.

If you want to send someone a file, such as a document, picture, or PDF, tap the Files button in the upper left corner, select File Share, and choose what you'd like to share.

Finally, if you don't want anyone else to be able to view your conversation history, check the "Only Me" option under Privacy. This setting prevents anyone else from seeing your previous chats.

Chapter Five: Taking Photos and Videos

The iPhone 12's new cameras make it simple to take photos by tapping once. The iPhone 12 features three cameras: a wide-angle lens, an ultra-wide-angle lens, and a telephoto lens.

There is no single "best" setting for capturing subjects outside the phone; each camera has advantages and disadvantages. Even if you don't own a DSLR, it's always a good idea to become acquainted with the camera you intend to use for your photos.

How to Take a Simple Photo

For starters, you can take a simple photo by following these simple steps:

1. Open the Camera app.
2. Select Photo mode
3. Face the Camera
4. Tap the Shutter

Basic Settings

To switch between taking pictures and recording videos, open the Camera app and tap the photo icon at the bottom of the screen. You can choose from the following options by swiping left or right:

- **Video:** Choose the video option and press the button to start recording a video.
- **Portrait Mode:** Select this mode if you want to photograph people outside. You can see the entire body of the person in front of you while keeping the background blurry. Zoom in closer than usual to ensure that the person does not appear too large.

- **Landscape Mode:** For landscapes, go into Landscape Mode. Zoom in close enough to keep the horizon level but far enough away so you don't lose detail.
- **Square Mode:** When taking pictures of objects, such as a building, use Square Mode. Use a wider field of view to include more of the object. Double-click the screen on iPhone 12 and later devices to open the camera app if you want to take photos with a square ratio. Then, select either "Square" or "4:3."
- **Time-Lapses:** Create a time-lapsed movie of moving objects. See time-lapsing videos.
- **Panorama:** Capture a panoramic landscape scene or other scenes; take a panoramic picture.
- **Cinematic:** Use a DoF (or similar) filter to add a cinematic look to your video recordings; see recorded Cinematic videos for examples.

If you want to change the photo app from which photos appear when you take them, go to Settings > Photos & Camera > [Camera] > [Save As].

Use iCloud

Your photos and videos are automatically synced across your Apple devices, iCloud.com, and your personal computer when you use iCloud Images. While using iCloud Photos, it is unnecessary to transfer photos from one iCloud-enabled device to another. You can be confident that iCloud Photos will only back up and keep the highest-quality versions of your photos. Original high-resolution files can be stored on your devices, or smaller, more space-efficient versions can be used. In any case, the originals can always be downloaded. Any documents you've edited or updated will always be reflected in the most recent version on your Apple devices.

Use Airdrop to Send and Receive Photos

To send and receive files between adjacent Apple devices, use AirDrop. But first, you have to ensure that the factors listed below have been met:

- Ensure the recipient is close by and within Bluetooth or Wi-Fi range when sending something.
- Make sure you and the person you're sending there are connected to the internet and Bluetooth enabled. Personal Hotspots should be disabled immediately.
- Ensure the person you are sending to has not restricted their AirDrop to accepting connections only from people in their contact list. If they do and you're already in their Contacts, AirDrop will only work if they have your email address or mobile number associated with your Apple ID in your contact card.
- To receive the file via AirDrop, even if you aren't in their Contacts, they will need to change their receiving option to Everyone.

To restrict access to your device and the files you receive via AirDrop, you can toggle the setting to "Contacts Only" or "Off" at any moment.

How to Use AirDrop

1. Open the Photos app (you can choose several photos to share from the app by swiping left or right.)
2. Tap on the Share menu or button.
3. Pick AirDrop from the list of options.
4. Choose the person you want to AirDrop to from the list. You can also use AirDrop to send files between your own Apple devices.

If your device's AirDrop button is red and has numbers, there is more than one device in range. When you click the AirDrop icon, a list of people who can receive your file will appear.

How to Receive AirDrop

If you receive an AirDrop, you'll receive a notification with a preview of the file. A tap on the Accept or Decline buttons may indicate a positive or negative response.

If you accept the AirDrop, the app that sent it will open. Images are displayed in the Photos app, and web pages are loaded in the Safari browser. When a link to an app is clicked, the App Store is launched.

Internal AirDrop, such as a photo from an iPhone to a Mac, does not display the Accept or Decline buttons; the item is sent immediately. Simply sign into both devices with the same Apple ID.

How to Get the Best Photos/Videos

In addition to the standard settings, you can access several choices, including Night Mode, HDR, Zooming, and Slow-Mo Motion. These options are readily available when you open the camera app.

Night Mode

The iPhone 12 has a brand-new night mode feature that allows you to capture stunning images in low-light conditions without a flash. If the iPhone detects a lack of ambient light, it will automatically extend the shutter for several seconds to ensure that the camera captures the appropriate amount of light.

When the feature is enabled, the iPhone displays a moon-shaped icon at the bottom of the screen to assist you in getting the best shots possible. When you tap the icon, the iPhone displays a slider where you can change the exposure settings.

The number next to the icon will change depending on how much time you give the phone to capture the photo. For example, if you specify 5 seconds, the number may appear as "5 sec," indicating that you want the camera to wait five seconds before taking the shot.

Simply swipe up from the bottom of the screen to reveal the full viewfinder if you want to see your current settings. You'll find the brightness level as well as the ISO setting there.

In addition to controlling the exposure, you can also control the focus. You can move the focus around the frame by tapping the arrow in the bottom left corner of the viewfinder. Once you've found the area you'd like to keep sharp, tap the button again to lock the focus where you want it.

HDR

High Dynamic Range (HDR) photography allows your smartphone to take multiple photos at different exposures, which can be combined into a single image with a much brighter and clearer image than a single shot would allow. This example video demonstrates how well it works.

To enable HDR (high dynamic range) photography, tap the HDR button at the top of the camera app. Then, using the shutter button, take photos. Your images will be converted to a high dynamic range format automatically. Tap the shutter button again to return to normal.

You can now take a photo that will appear brighter than a regular one. Because it takes multiple photos, it creates better contrast - even in low-light situations. And because it captures multiple shots, it creates better contrast — even in low-light situations.

Zoom in to Take Photos of Objects Far Away

Open the camera app, tap the little gear icon next to the shutter button, and select "Photo." This brings up the camera interface, where you can choose five modes. The Photo mode is selected by default, indicating that you are ready to take photos.

You can zoom in and out to change the image size or zoom in closer to see individual pixels. Simply drag your finger across the screen to accomplish this. You can also zoom in and out by touching the screen's corners.

You can also adjust the amount of digital zoom used on the image. To access the settings menu, tap the screen's bottom left corner. You can adjust the amount of zoom applied to the image by tapping each option.

Tap anywhere else on the screen to take a photo without zooming in. Your camera will still focus and capture the shot automatically.

Slow-Mo Mode

Slow-Mo mode allows you to record life in slow motion. The feature allows you to record special moments such as a car accident, a fight scene, or a wedding proposal. You can use the camera app's Slow-Mo mode to freeze motion for seconds, allowing you to capture those rare occurrences.

1. To activate the mode, swipe the options below the shutter and click on slow-mo.
2. Adjust the speed to your preference by dragging your finger on the adjustment scale.
3. Once you get your preferred speed, tap the shutter button and hold to record a clip.
4. To adjust how quickly the view zooms out, tap the rotation control in the bottom left corner. You can set the view to the entire camera or the front or rear lens. By default, the view is set to the entire camera.

You can snap photos at 120 fps, 180 fps, or 240 fps. If you want to shoot videos at the same speed, keep the shutter taped longer. When you're done recording, tap the volume buttons to release the shutter.

What's Up with the New Camera Lenses?

While the iPhone 12 Pro and Pro Max have three cameras on the back, the regular iPhone 12 and mini have only two cameras, each with a unique perspective. You can use a standard zoom lens or an ultra-wide or wider-angle lens.

Wide-angle lenses highlight what's happening in the background when photographing a subject from an extremely close distance. However, when shooting landscapes or cities, use a wide-angle lens for maximum coverage.

You can adjust the focal length in both options by pinching the screen and sliding an adjustable slider. The table mode allows you to zoom in up to 0.5x.

Chapter Six: Apps for Seniors

You might be looking for new iPhone apps daily. Nonetheless, here is a short list of apps that may whet your appetite and interest you in exploring everything the iPhone offers. Accessing the app store is simple; the iPhone's Home screen includes a shortcut to the App Store. If you still don't know how to access the app store, read chapter three for the complete guide.

To begin, go to the app store and select either the Today tab (which highlights featured apps and content) or the Categories or Top Charts tabs (see the buttons at the bottom of the screen). You can also look for programs by using the Search button. To learn more about a certain app, click on it.

Now that you can navigate the app store, let us go through some applications that would make your iPhone more efficient, fun, and helpful.

Games

Sudoku

A fun brain exercise that most elderly people know well. It has three lessons, each with several levels, which makes it an excellent way to pass the hours in a doctor's or dentist's office. Several Sudoku apps are available, so pick the free ones and the best fit for you.

WordBrain

WordBrain is a word puzzle game designed to improve memory and cognitive skills. Players are presented with words that must be combined into longer phrases. Words are arranged in three or four groups, each representing a letter. As you progress through the levels, the number of letters required to complete each phrase increases.

The game features both single-player and multiplayer modes. In single-player mode, players compete against themselves to see how many points they can earn. Multiplayer allows up to eight people to play together simultaneously.

Candy Crush

This game is addictive and easy to play. The object is to match candies by swiping them into place. Candy Crush is free, and there are many versions available as well.

Health and Fitness

Epicurious

Eating good food is essential to staying healthy. If you're into cooking and want inspiration from thousands of delicious recipes, Epicurious is an iPhone app. You don't need to flip through a cookery magazine to get ideas.

MyFitnessPal

MyFitnessPal is one of the most popular apps for tracking your diet and exercise. With it, you can keep tabs on what you're eating and how much you're moving and connect with others trying to live healthier lives.

Nike Training Club (Free)

Create customized routines, watch video demonstrations of new exercises, and follow step-by-step instructions with this useful tool. The incentive scheme of this iPhone app can help you stick to your exercise plans.

Productivity

Goodreads

Goodreads is an excellent way for readers to stay updated with new books and authors. You can create personal reading lists and read book recommendations from others.

LibriVox

This app allows you to download audio versions of public domain books for listening on the go. You can stream the book directly from the app or save it to your device for offline playback. There are over 27,000 books available for download. New books are added daily, and there are even new genres like children's books, science fiction, and fantasy.

Pigment – Adult Coloring Book

Over the years, the popularity of adult coloring books has skyrocketed, and they've shown to be a wonderful tool for relaxing and letting our creativity flow. The top coloring book applications for the iPhone include this one.

Audible Audiobooks & Originals

Audible allows you to buy audiobooks online or download them to your device. Audible allows you to read eBooks, magazines, newspapers, and comic books. Or perhaps you want to unwind by reading fiction or nonfiction. Listening to audiobooks while commuting, exercising, cooking, cleaning, or doing anything else is possible. And because it's digital, you can take your books wherever you go. This is especially beneficial for seniors who struggle to read large amounts of text or any text at all.

Finance

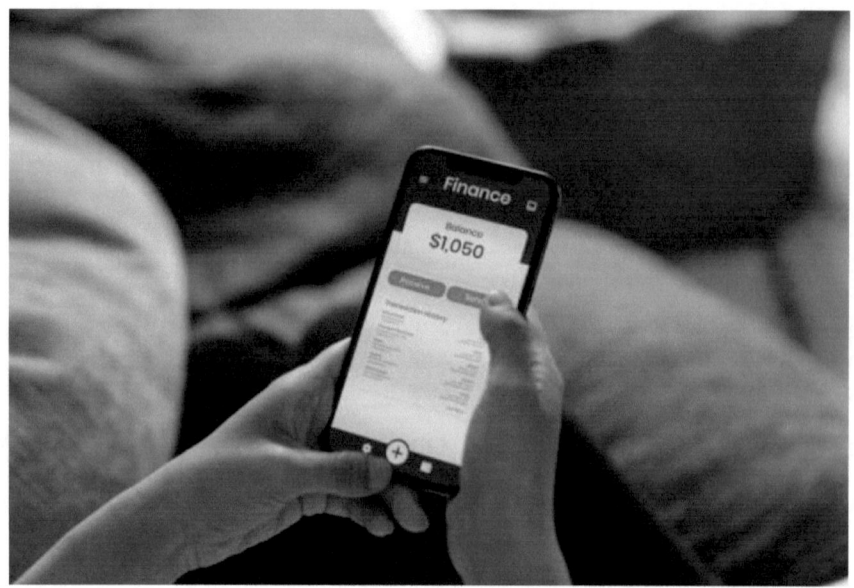

Mint

With the help of this highly-rated financial software, you can manage all of your accounts, which have received over 400,000 positive reviews on the App Store and a near-perfect rating of 5 stars. View account balances, track payments and spending, create budgets, and more!

Bank of America Mobile Banking

With Bank of America mobile banking, you can access your checking, savings, credit card, debit card, and mortgage accounts right from your phone. You can also check your balance, pay bills, transfer money between accounts, and more. It's a great way to monitor your finances and stay organized.

PayPal

This secure payment service makes sending money as easy as paying for things through email. You can send money to anyone in the world securely and instantly. No bank account is needed.

Medications

GoodRx

This iPhone app is a goldmine for finding the best local prescription prices. Instead of going to the same old drugstore and hoping to get the best deal, it has saved users thousands of dollars. You won't have to cut coupons from flyers or newspaper ads; GoodRx will find them for you!

Blood Pressure Monitor

With the help of this iPhone software, you may monitor your blood pressure and store data for a long time in one handy location on your iPhone. Using the included reports, give your doctor a thorough understanding of your blood pressure.

Fitbit

The Fitbit is a wearable fitness tracker that tracks steps taken, distance traveled, calories burned, floors climbed, sleep quality, and other metrics. It also provides feedback on how active you are throughout the day. It works with iOS devices only.

MoodFit

MoodFit is a free mobile application that helps people track their moods over time. With it, you can see how your mood changes throughout the day and week and gain insights into what affects your emotions. You can use the app to monitor your moods and identify triggers that affect your emotional state.

Medisafe Medication Management

This app helps you remember to take your medication, even if your phone is asleep. It automatically checks your prescription bottles every morning and evening and alerts you if you forget to take your pills. You can try out the free version if you don't want to pay for the full version.

Communication

WhatsApp

WhatsApp is one of the most popular apps out there. It allows you to chat with your friends and family without paying SMS fees. It even lets you do video calling over Wi-Fi. You can check chapter 3 for further information.

Skype

Free online calls can be made to friends and relatives. Even if your iPhone has FaceTime, some people in your circle might not have an iPhone; therefore, Skype would be the ideal option in those cases.

Senior MeetMe

The app is specifically designed for adults aged 45 and up and limits access to verified members 55 and older. You'll see photos of other members and information such as age, location, occupation, relationship status, and hobbies.

You can browse based on interest categories, including travel, sports, music, movies, food, pets, culture, religion, politics, and even spirituality. And unlike many online dating sites, you don't have to pay anything upfront to use Senior MeetMe.

Entertainment

Netflix

The Netflix application provides access to thousands of movies and TV shows that you can watch instantly on your iPhone. With its vast content library, including documentaries, TV shows, and hundreds of movies, you'll never run out of things to watch again.

Pandora Radio

Pandora is a free music streaming service that offers personalized radio stations based on your favorite artists and songs. You can listen to millions of tracks, including indie rock, hip-hop, country, jazz, and classical music.

YouTube

If you're looking for videos about health, finance, fitness, or anything you are curious to learn about, YouTube is the place to go. There's no shortage of videos available for you to view. You can search for topics like "how to lose weight," "best exercises for seniors," and more.

Chapter Seven: Siri

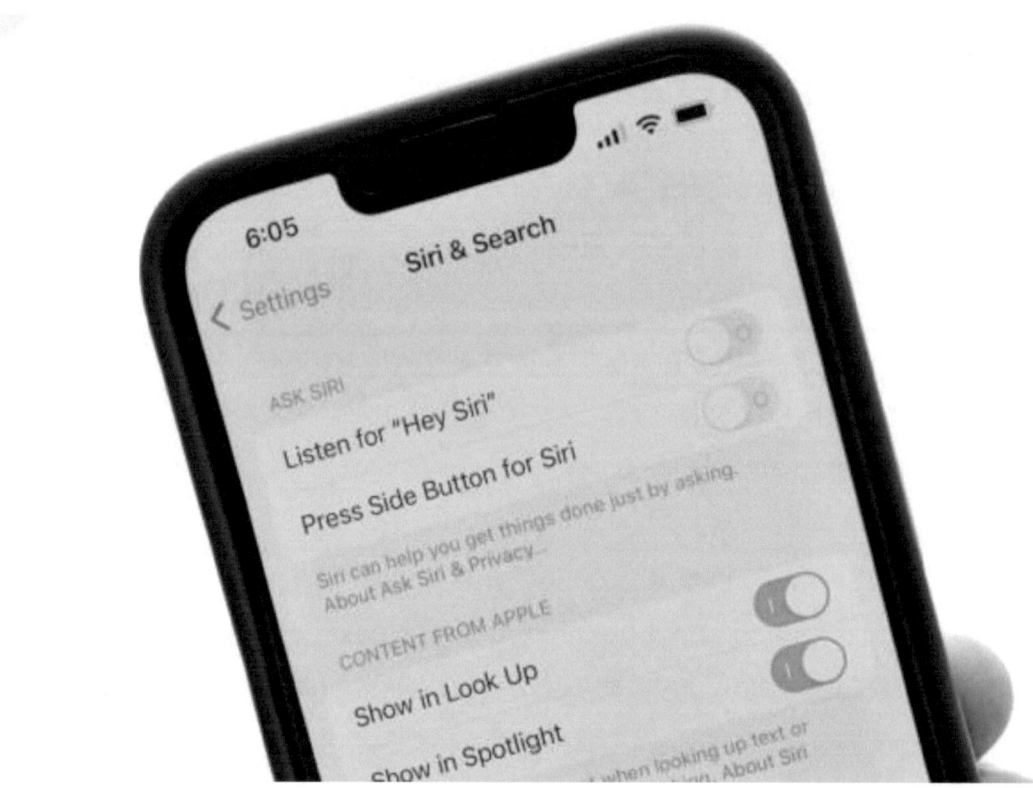

Siri is a virtual personal assistant who can assist you with various tasks such as making phone calls, sending messages, finding directions, setting reminders, etc. It operates by listening to your voice commands.

Siri is the best tool for seniors with an iPhone 12; it is excellent at answering questions about sports scores, weather, stock prices, movie times, etc. Ask Siri to read an article from The New York Times or The Wall Street Journal aloud. Siri can also be used to get directions. Siri can also make phone calls, send text messages, locate locations, play music, control smart home appliances, and much more.

How to Set Up Siri

Setting up Siri isn't complicated; you can get your virtual assistant up and running with these three steps:

1. Look for and tap the "Settings" icon on the iPhone's home screen.

2. Select "Siri & Search."

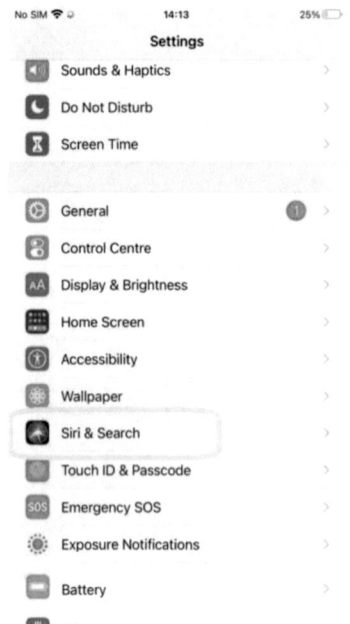

3. Select "Listen for "Hey Siri."

This will open a Siri setup page, where the user will be prompted to say a few commands to Siri so that it can learn their voice and get to know them. Seniors should ideally perform this task to learn which commands Siri can recognize.

When this feature is activated, you can show the elderly person how to say questions like "Hey Siri, what's the news?" and "Hey Siri, what time is it?" as well as "Hey Siri, call Jonathan."

How Siri Can Help You

Siri can be programmed to help you simplify your life and tasks. The software's potential applications include requesting the time, date, news, weather, future events, market prices, and sports scores. It also functions as a contact manager, instant messenger, and email client. The system can even handle lighthearted interrogation.

It can converse with humans not only in English but also in several other languages. This software works primarily as a voice-activated internet search engine. In this section, we'll go over some of the smartest things Siri can do for you.

Seek Siri's Help to Set Alarms and Reminders, Add to the Calendar, and More

To get Siri to do things for you, you first have to say, "Hey Siri," and then speak any of the following commands:

- "Put the alarm on for 15 minutes."
- "Start the clock over."
- "It's time to put the timer on hold."
- "Exactly what time is it?"
- "Can you tell me the date today?"
- "Get me up in an hour."
- "Remember to wake up at 7 a.m."
- "Silence all warning devices."
- "I need a gentle nudge to remind me that it's time to take my pills at eight o'clock."
- "Set a two-minute timer."
- "Remind me at 6 PM to call Joan."
- "Set the alarm for 5:30 AM tomorrow."
- "Don't forget to tell me to call the dentist when I get home."
- "Send Don Frederiksen a text message." (She'll ask what you want to say).
- "I want to ensure I remember to send Suzie a birthday card the following Wednesday."
- "Schedule a trip to the dentist for next Tuesday at 10 in the morning."
- "Please schedule me for iPad instruction next Tuesday at 1 pm."
- "Get the iPad class off my schedule."

Command Siri to Check and Reply to Messages and Emails

- "Check my inbox."
- "Send an email to Justin,"
- "Please tell Joe how much I appreciate all his hard work over the weekend by emailing him."
- "Tell Pat to wait for me by texting him on his cell phone."

- "Please peruse my most recent email."
- "Check out my new text messages."
- You can also ask Siri to read aloud the sender's name, the email's subject, and the time it was sent.

Produce a Written Grocery List (Substitute the Desired Name of Your List)

- "Make a note to get bananas, milk, and bread at the supermarket."
- "Get the note with the grocery list."

Educative Purposes

- "Snap a Pic."
- "Can you tell me how old Tony Bennett is?'
- "In other words, how much would a tip of 15% on $9.99 be?"
- "The trip home, how long does it take?"
- "How is the weather going to be today/tomorrow?"
- "What coffee shop can I visit?"
- "The TCF Bank Stadium can be easily reached by following these directions."
- "Contact Joe Friendly on his cell phone. (Joe's cell phone number, or mobile, rather than his landline)."

Command Siri to Play Music or Radio or to Locate Specific Apps

Substitute the title of a song, an artist's name, or an app's name if you want to listen to something from your iPad's music library.

- To Sing "Amazing Grace."
- It's time for some tunes, so turn on the radio."
- "Play some Michael Jackson music."
- "Repeat songs from my top playlist" (substitute the name of a playlist located on your iPad)
- Play solitaire while the song plays by pausing it or skipping it.

Ask Siri to Launch Apps

- "Launch App Store"
- "Open Magnifying Glass,"

For Fun

Siri is also a great way to relax and have a good time. Have a go with them.

- "Just what is the point of living?"
- "Which comes first, the chicken or the egg?"
- "Siri, please tell a joke."
- "Knock, knock."
- If you divide zero by zero, you get nothing.

Siri Can Do These Other Things When You Tell Her to

- "How hot or cold is it today?"
- "Will Phoenix get blisteringly hot this weekend?"
- "What does tonight's weather seem like?"
- For those with diabetes, try looking up pasta dishes.
- Look up the latest D.C. news online.
- Google for "dog pictures."
- "What movies are showing in my area?"
- "Who played the lead role in Casablanca?"
- "Please display Sunday's football standings."
- "Is it possible to count down the days till Christmas?"
- "Where can I find Mexican food near me?"
- "Can you tell me the conversion from cups to pints?"
- Say to Siri, "Note: the car is parked in Row 5, Level 3, at the airport." Siri will then create a note and locate it for you.

Fun Facts to Know About Siri

- **Siri is convenient and useful:** You might be surprised to learn she has saved lives. For example, Forbes reported that someone in a dangerous situation could use Siri's voice commands to call for help.

- **Siri takes up no screen space:** With the most recent version of iOS (iOS 14), the voice assistant now appears as an icon in the lower right corner of the iPhone's display. When you ask questions or make requests, the responses appear as small icons and advertisements in a portion of the display but not the entire display.

Chapter Eight: Best Tips and Tricks with iPhone 12

How to Make the iPhone Easier for the Elderly

Apple products are fantastic because they are simple to use. With a few simple changes, any iPhone can be transformed into an excellent phone for the elderly. If you want to give an iPhone to elderly parents or friends and ensure the device is appropriate for their needs, you've come to the right place.

Enlarge the Font for Easier Reading

Elderly people frequently have visual problems, but fortunately, we can increase the tiny text appearing in the iPhone menus and messages. This is how you can make the text on your iPhone larger:

1. The Settings icon should be located on the iPhone's home screen. Tap it.
2. Select "Accessibility."
3. Select "Display & Text Size."
4. Select "Larger Text."
5. Increase the font size by dragging the slider to the right at the bottom of the screen.

Show Them How to Utilize Siri

Natural speech is the most natural way for elderly people to interact with technology. By learning to use Siri, the elderly person can simply make calls, read their daily news aloud, and get answers to questions. To get Siri to answer questions at any time, follow the step-by-step instructions in Chapter 6.

Select "Spoken Content" from the Menu

The ability to read the screen's content aloud on iPhones can be useful for people who are visually impaired. This is especially useful if the user wishes to browse the internet and enjoy written content, such as Wikipedia articles, without having to read those materials independently. Follow these steps to enable "Spoken Content":

1. The Settings icon should be located on the iPhone's home screen. Tap it.
2. Select "Accessibility."
3. Select "Spoken Content"
4. To activate it, tap "Speak Screen."

You can slow down the speech pace by dragging a slider at the bottom of the same screen.

Swiping down with two fingers from the top of the screen causes the iPhone to read aloud whatever is on the screen. Ascertain that the user understands how to use the gesture and provide written instructions if necessary. This function is best used in text messages, web pages, and news stories.

Increase the iPhone's Ring Volume

You can help someone who has trouble hearing by turning up the iPhone's ringer volume. To increase the ringer's volume, you can do a few things:

1. Locate the "Settings" icon on your iPhone's main menu and tap it.
2. Choose the "Audio & Touch" option.
3. Raise the "Ringer and Alerts" bar to the right.
4. The slider's rightward movement increases the volume of the phone's ringtone. To change the iPhone's default ringtone, tap the "Ringtone" option.

Establish "Emergency SOS" and Instruct on Its Use

The "Emergency SOS" function immediately contacts the appropriate authorities and notifies the user's pre-programmed emergency contacts. On older models, the user must quickly tap the power button five times, while on models with iOS 8 or later, the user must tap and hold the power button and one of the volume keys. For more information and to add yourself as an emergency contact, follow these steps:

1. On your iPhone's main menu, locate and select "Settings."
2. Select the "SOS" option if you need urgent help.
3. To configure your emergency contacts under Health, select "Auto Call."
4. To make changes, click "Edit."

If "Emergency SOS" is enabled, you can enter the senior's name, age, medical conditions, location, and any other information that emergency services may require.

If the individual has activated "Emergency SOS," you can be contacted by tapping the "add emergency contact" button at the bottom of the page. Because the exact nature of this feature and how it works on different iPhone models and in different locations can vary, you should contact Apple support or visit your local Apple Store for assistance.

Activate "Find my iPhone."

If the iPhone ever goes missing, or if you just need to keep tabs on the user's whereabouts, turning on "Find my iPhone" is a great idea. This feature can be activated by:

1. On your iPhone's main menu, locate and select "Settings."
2. Use the image above to tap the user's name.
3. Try the "Find My" button.
4. Click the "Lost iPhone" button.
5. Allow all three choices to be active.

If you and the elderly person are members of the same Family Sharing group or use the same Apple ID, you can use your respective "Find My" apps with this feature enabled to track each other's whereabouts.

How to Use Apple Maps

Now is the time if you haven't already become acquainted with the built-in maps application on your iOS device. The Apple Maps application comes preinstalled on all Apple devices. It provides directions and information about nearby businesses, restaurants, and attractions, among other things. You can see how crowded the beach is, the weather forecast, and the location of the nearest gas station. You'll also be able to see local landmarks before you arrive if you're planning a new trip.

Apple Maps can help you both plan where to go and arrive safely. Whether you're picking up the kids or meeting friends for dinner – this app has you covered. Here we've got a step-by-step guide to getting the most out of this app.

Let's go over how to use Google Maps to get directions from one location to another, share our current location, and configure Apple Maps so that we can use it most effectively.

Getting Familiar with Apple Maps

We'll review how to use Apple Maps on your iPhone and become acquainted with its features. If you don't know where the app is, you should look for it first. You can try searching for the application if you cannot locate it.

The app can be launched in several ways:

- To use it, simply press the app icon on your home screen.
- Swipe right from your iPhone's home screen to access Apple Maps. You may need to scroll to the bottom to see the View button. To open Apple Maps, swipe right on your home screen. To open Apple Maps on your iPhone, swipe right from the bottom edge of the screen.
- If you click an address from any app, Apple should automatically open maps so you can get driving instructions there.

Add Important Addresses

Once you've opened the Directions application, you'll see a list of "Favorites." You can add your home location and any other locations you use frequently, so you don't need to enter them every trip.

1. Press the "+" icon in the bottom right corner of the screen to add a new location.
2. Click "Home," "Work," or "Add." Viewing your favorites
3. Adding an address to your favorites

It's simple to add addresses like your friend's house or your regular grocery store. Simply enter the location you want to add and click "done" in the right-hand corner when finished.

Choose Your View: Transit, Map, or Satellite

To change the map's views, click the button in the bottom right corner of the screen. For a satellite view, a bus/train view, or walking directions, select "Trails" from the menu.

On the top right part of the screen, tap the "i" icon. Furthermore, tap any of the three options you'd like to use:

- Map view
- Transit view
- Satellite view

Pin, and Share Your Location

You can use the map tool to see where you are or share your current position with others by using the Share button.

Tap the "i" icon. Then, tap "Mark my Location." This will create a red pin marking the location that you tapped.

Mark Location

Hit the "Share" icon. This will allow you to message your location to someone else.

Share a Pin

If you don't want that particular place, tap the red pin to remove it from your map.

How to Preserve Battery Life/Battery Save

Numerous iPhone users have expressed dissatisfaction with the device's short battery life. Many people struggle just to get by each day. Users are concerned about the iPhone 12's quick battery drain and battery life, as they are with every new operating system release.

Various factors, including excessive use of GPS, system apps and games, and other user actions, can cause battery life issues with new iPhones. While you can't do anything about a bug draining your battery life until Apple fixes it, you can take steps to improve your battery life and reduce any hidden sources of drain.

1. Control How Often and When Apps Can Access Your Location

Battery life and privacy are two solid reasons to review your location settings and reduce the number of apps that have access to your location. Follow these steps to access your device's Location Services menu:

1. Open the Settings app.
2. Click on Privacy.
3. Tap on Location Services.
4. Simply make changes by taping the app's name in the list to disable its access to the location.

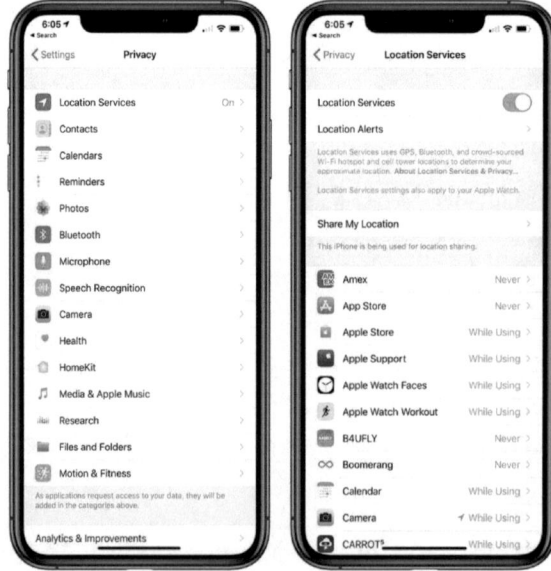

There are four distinct options for app-specific location settings, albeit not all four may always be accessible based on the app's specific functionality. There are a few options: Never, Ask Next Time, In-App, and Always. Here's how these options work:

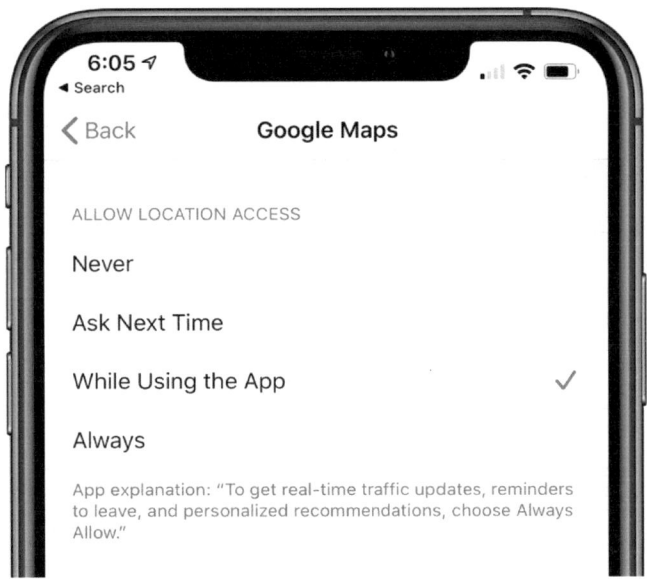

1. The best choice, unless required, like a mapping app, is to set location access to Never. This will prevent any app from ever accessing your location.

2. If you choose to Ask Next Time, the app will display a pop-up asking permission to access your location the next time it needs it. This option disables location services until the user manually activates them using the pop-up window.

3. While Using the App, setting limits location detection to when the app is being utilized. When you exit the app or go to another app, location services are disabled.

4. Lastly, giving an app always access to your location means that it can track your movements while the app is open or closed. This is the worst battery life, so only use it for your needed apps.

Removing junk from this section ensures that no apps use your location without your permission. For example, a banking app may request location access to display nearby ATMs, but the same information can be obtained by entering a zip code.

Even though it is possible to disable Location Services completely, only a few people do so because it can cause issues with useful apps such as Maps.

2. Limit the Number of Apps Using Bluetooth

A surprising number of apps attempt to use Bluetooth for location monitoring or scans for Chromecast devices, and the iPhone 12 includes a feature that allows you to see when apps request Bluetooth access.

Checking this list will prevent your battery from being drained by an app secretly making Bluetooth connections in the background without your knowledge. Apps that use Bluetooth-enabled accessories should have access to the feature, but retail stores should not. To change your Bluetooth preferences, follow these steps:

1. Open the app for Settings.
2. Click Privacy.
3. Select Bluetooth.

Apps on this list that do not require Bluetooth should be disabled. Minimizing access is the best course of action. If you disable Bluetooth and a feature in an app stops working, turn it back on.

You can also disable Bluetooth entirely, which may save some battery life but isn't a good idea for most people because Apple Watches, AirPods, and other accessories use Bluetooth.

3. Use Wi-Fi More Often

Connecting to Wi-Fi rather than cellular data saves battery life, and Apple recommends doing so whenever possible. Turning on Wi-Fi at home or work will keep your phone's data plan and battery from being depleted.

4. Use Low Power Mode

Regarding battery life, the most important setting to enable is the long-standing Low Power Mode. It slows down background downloads and accelerates the rate at which your screen's brightness dims after you've been inactive for a while.

When your iPhone's battery reaches 20%, a prompt appears to enable Low Power Mode, but you can also do so at any time by touching the battery icon in Control Center or asking Siri to do so. You can also access it from the Settings menu:

1. Go to Settings.
2. Go down and select the Battery option.
3. Turn On Low Power Mode.

The battery symbol on your iPhone will turn yellow to indicate that Low Power Mode is active. Certain users prefer Low Power Mode, which should be kept active at all times;

however, it must be manually enabled because it is disabled whenever the iPhone is charged.

5. When in a Weak Signal Area, Go To Airplane Mode

When searching for a signal or connecting in an area with poor or no cell phone service, your iPhone consumes more power than is necessary. Because you won't be able to do anything else if your cellular reception is poor, you can choose to use Airplane Mode.

6. Verify the Condition of Your Battery

An aging and inefficient battery may be the cause of battery drainage. These procedures will help you determine if your battery needs replacing.

1. Go to the Settings app
2. Go down and select the Battery option.
3. Choose Battery Health from the menu.

One of the metrics displayed in the Battery Health section is the maximum capacity, expressed as a percentage of the battery's original capacity when it was brand new.

The battery should be replaced if the current capacity is less than 80%. If your battery capacity falls below 80% while your device is still under warranty or AppleCare+, Apple will replace it for free.

If your iPhone's battery dies, you can expect to pay between $49 and $69 to replace it. You can help your iPhone's battery last as long as possible between charges by enabling Optimized Battery Charging in the Settings app's Battery Health section. When you charge your iPhone, it remembers when you last charged it and will only charge it to 80%.

If you charge your iPhone overnight, the Optimized Battery Charging setting may keep it at 80% charge until you wake up, delaying the battery's aging process. To extend your battery life, Apple recommends taking precautions such as keeping your device out of extreme heat or cold and removing certain cases when charging. If the iPhone case is too hot to charge, remove it to extend the battery's life.

7. Limit Apps Running On Background

When you turn off the background app refreshing, your apps will no longer load your mail or download updates in the background while you're using them.

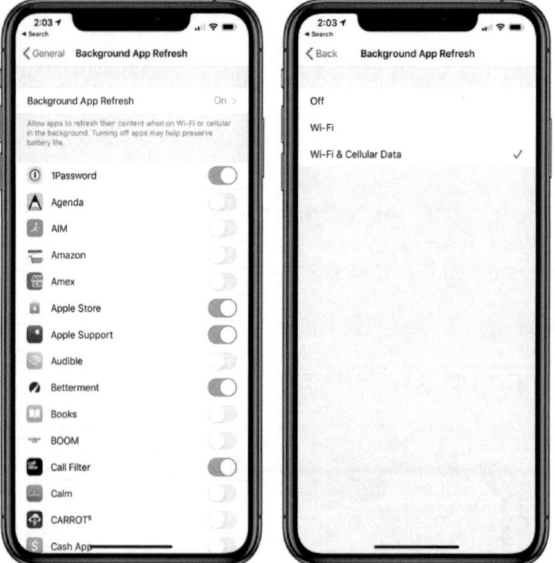

Disabling Background App Refresh may help you get more out of your battery. Background App Reload can be turned off completely or selectively, allowing only specific apps to refresh in the background.

1. Launch the device's configuration menu.
2. Choose the default setting, General.
3. Select the option to refresh background apps.

You can disable Background App Refresh completely by tapping the option again, or you can set it to work only when connected to WiFi, which is more energy-efficient than cellular data.

Toggle the switch next to each app to enable Background App Refresh only for the ones you use the most.

8. Manage Power-Sapping Apps

The iPhone can now tell you which apps are draining your battery, making it much easier to ensure that nothing is draining your battery covertly and without your knowledge. To view information about your device's battery usage, head to Settings and navigate to the Battery section.

You can view a graph of your battery life over the last 24 or 10 days and a list of the apps that have used the most power. If you discover that a program consumes too much power but is no longer in use, you can uninstall it.

Even if you need the app, you can save battery life by limiting its use. The duration of an app's use of Background App Refresh will also be specified here.

9. Modify Your Mail Fetch Settings

Using the Mail app, you may conserve battery life by customizing how often it checks for new emails and turning off Background Refresh.

1. Go to settings
2. Select Email,
3. Select Accounts.
4. Choose the option "Fetch New Data" below.

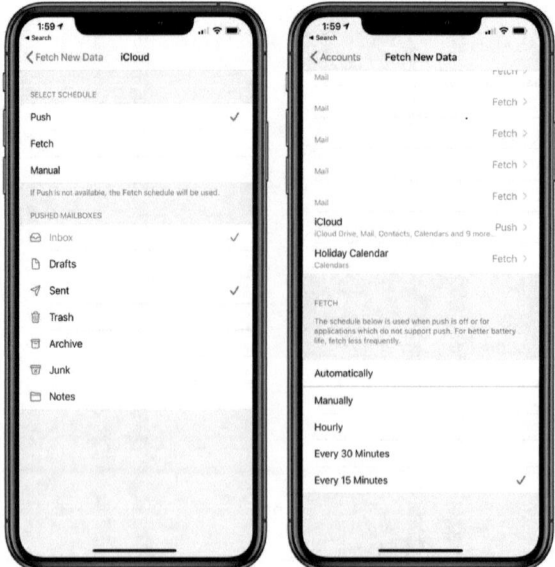

If your email provider does not support Push notifications for new messages, you can disable this feature and configure your account's Fetch settings here (like Gmail accounts).

To save battery life, increase the Fetch check interval or disable Fetch entirely and perform manual checks to download new messages only when the Mail app is launched. Your options are manual, Automatic, Hourly, Every 30 Minutes, and Every 15 minutes.

10. Reduce Notifications

Limiting the number of alerts sent by your apps can help you save power. Not only is being constantly reminded of new messages and alerts from your various apps annoying, but it also drains your battery life.

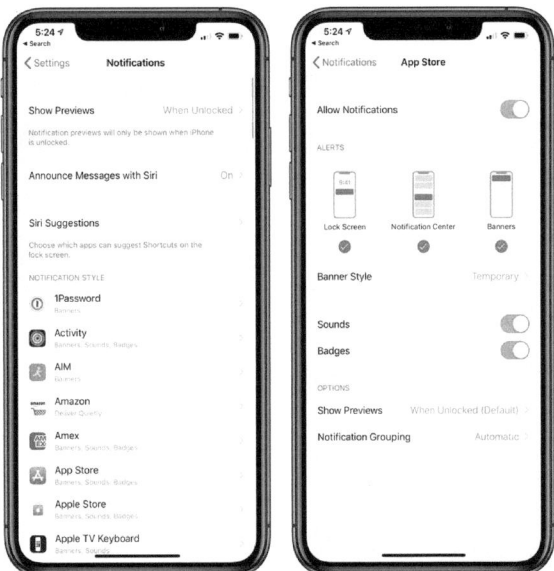

Follow these steps to modify your alert preferences in the Settings menu:

1. Go to Settings.
2. Select the Notifications option.
3. Using your device, you may go through each app and set the "Notifications Enabled" switch to control which apps will send you alerts.

Alerts can appear in three places, depending on your preferences: the Lock screen, the Notification Center, or as banners.

Apple includes a convenient feature that allows you to change your notification settings directly from the Lock screen alert. Long-tap it to silence or dismiss a notification and then tap the three dots (...). In addition to completely turning off notifications for that app, you can have them delivered quietly to Notification Center but not to the Lock screen.

Furthermore, the Focus Mode can silence messages for most of the day. The Notification Summary in iOS 15 is an excellent tool for managing the constant stream of alerts. The system compiles your alerts daily and sends them to you at a predetermined time. While it is still recommended that you disable as many unnecessary alerts as possible, Notification Summary can assist you in receiving the notifications you require without missing a beat.

11. Stop All Future App Updates and Downloads Automatically

If the battery life on your iPhone is low, you probably don't want it to download apps from the App Store or update its software without your knowledge.

Apple's iCloud service can automatically update your devices with the most recent versions of any programs you've installed on any of them. Any app installed on an iOS device, such as an iPad, is automatically downloaded and installed on any iOS device you sync. Leave that option active if you want it, or disable it if you don't.

1. Open Settings App
2. Select your image from the menu.
3. Select the App Store or iTunes menu.
4. Toggle off the playback of music, apps, and books/audiobooks.

If you do not want your apps to be automatically updated, disable App Updates. If you keep this toggle turned on, your iPhone will automatically update whenever an app receives a new version from the App Store. If you do not want your iOS device to automatically update, you can disable it by doing the following:

1. Click on Settings
2. Select the Menu.

3. Use the menu option to upgrade your software.

4. Select the Auto-Update option.

5. To disable updates, simply tap the toggle.

Maintain and Clean Your iPhone 12

The iPhone 12 models' back glass and camera section are made of glass with a sophisticated textured matte finish. Normal wear and tear on your iPhone may result in material transfer from nearby jeans or other pocket contents on the textured glass. Material transfer can appear as a scratch, but it is usually simple to repair.

Clean it immediately if your iPhone comes into contact with anything that could stain or damage it, such as dirt, dust, ink, cosmetics, soap, detergent, acids or acidic juices, or lotions. Please clean according to these guidelines:

- Disconnect your iPhone from all devices and turn it off.
- Try a lint-free, microfiber cloth that you can slightly dampen (a lens cloth, for instance).
- Keep moisture away from cracks and crevices.
- Avoid using any type of cleaning solution or air conditioner.

The iPhone has an oleophobic coating, which makes it resistant to oily fingerprints. Abrasive materials and cleaning solutions can damage and even scratch the iPhone's protective layer.

Chapter Nine: FAQ

1. How many kinds of iPhone 12 are there?

Nowadays, Apple's flagships come in four different styles: the iPhone 12, the iPhone 12 Pro, the iPhone 12 Max, and the iPhone 12 Plus.

2. How much will an iPhone 12 cost?

With a base price of $999, Apple's new iPhone 12 Mini is the cheapest in the iPhone 12 lineup. All four models start at $999, but their pricing varies.

3. Is the Screen of the iPhone 12 Mini 5.4 inches?

It's the first time Apple has given one of its iPhones the designation "mini," and the new 5.4-inch iPhone 12 Mini, according to Apple, is the "smallest and lightest 5G phone in the world."

4. What are the iPhone 12 Mini's features?

The iPhone 12 Mini comes in two sizes — 6.5 inches and 7.3 inches. Both phones have the same display size, processor, and cameras as the larger iPhones. They also share the same design, including flat metal sides. However, they're smaller than the iPhone XS Max, which measures 8.2 inches diagonally.

5. What's different about the iPhone 12?

The iPhone 12 has an AMOLED screen, as did the iPhone 11 Pro Max last year. However, instead of an LCD panel, it has an OLED panel. And, unlike previous iPhones, it lacks a home button in favor of facial recognition technology, Face ID, while the camera setup remains unchanged. So, what does the iPhone 12 have to offer? Aside from making video calls via FaceTime, it can benefit from the faster speeds of 5GHz Wi-Fi networks. You'll also have longer battery life.

6. Is the iPhone 12 Pro Max better than the iPhone 12 Pro?

The iPhone 12 Pro has a larger screen than previous iPhones, but the price is comparable to previous models. It employs 5G technology, which enables faster downloads and uploads. Its camera has a new sensor that improves the sharpness and clarity of photos. It also includes a LiDAR scanner, which generates a detailed 3D space map.

On the one hand, the iPhone 12 Plus has a larger screen than the iPhone 11 Pro Max, measuring 6.7 inches. The iPhone 11 Pro Max has a smaller screen but offers superior image quality thanks to its Super Retina XDR OLED displays. Both phones have wider and ultra-wide cameras, larger sensor sizes, and 4K video support. They also have better audio recording capabilities.

How you use your smartphone will determine how you proceed. If you want a camera-focused device, go with the iPhone 12 Pro. If you want a phone that can do it all, including gaming, go with the standard iPhone 12.

The iPhone 12 Pro range is for people who love photography and have a lot of money. Apple is aiming the Pro models at photographers who take their work seriously.

7. Does iPhone 12 use IOS 14?

Yes, iOS 14 comes with all four iPhone 12 models.

8. Is the iPhone 12 safe from water and dust?

All four models of the iPhone XS Max have an IP68 rating, which means they're waterproof and dustproof.

9. Does the iPhone 12 have a 3.5mm jack for headphones?

No.

10. What color choices do the new iPhone 12 models have?

Apple has revealed the new iPhones for 2019. The iPhone 12 will be available in five colors: White, Red, Blue, Green, and Black. The iPhone 12 Pro, on the other hand, will be available in four colors: Graphite Silver, Space Gray, Rose Pink, and Midnight Blue. And finally, the iPhone 12 Max will be available in only one color: Space Gray.

11. Do the iPhone 12's charger and headphones come with it?

Apple has removed the earbuds and charger adapter from the iPhone 12 box. Apple claims to eliminate these accessories to reduce carbon emissions, stop mining, and make the iPhone less bulky.

12. Is there 5G on the iPhone 12?

Yes, all four models of the iPhone 12 have 5G. All four iPhones used in India can connect to Sub6.

13. But you're saying there's no new USB-C?

The new iPhone 12 models continue to use Lightning rather than USB-C.

14. How much storage space does the iPhone 12 come in?

If you choose the iPhone 12 or the iPhone 12 Mini, you can set it up with 64GB, 128GB, or 256GB of storage on the inside. The Pro line will have storage capacities from 128GB to 512 GB.

15. Does MagSafe work with all iPhone 12 models?

All four models of the iPhone work with MagSafe, so they can be used with any accessory that uses magnetic connectors.

16. Does spatial audio work on the iPhone 12?

All new iPhones will include spatial sound technology, making listening to music more immersive.

17. Can I use my old iPhone charger with the iPhone 12?

Your old iPhone charger will work with the new iPhone 12 models; however, if you want to charge them quickly, you will need to purchase a new charger.

18. Will the iPhone 12 work with AirPods?

Yes, the iPhone 12 works with AirPods. You can pair them using Bluetooth or by connecting them directly to the phone via a Lightning cable.

19. Does the iPhone 12 Support Dual-Sim?

Unfortunately, eSIM support is still patchy across carriers, so most of us will be unable to use this feature. On the other hand, the iPhone 12/12Pro lineup accepts one traditional SIM card and one eSIM card.

Besides China, Hong Kong, and Macao, no other region supports dual-sim functionality on the iPhone 12s. You can achieve dual-sim capability in these areas by placing one sim on top of the sim tray and the other underneath.

20. Does the iPhone have a headphone jack?

No, Apple hasn't allowed phones to have 3.5mm connectors for headphones in a long time.

21. Is a microSD card slot available on the Apple iPhone 12?

iPhones have never included them. The iPhone supports Micro SD and SD cards, but you'll need a compatible adaptor.

Conclusion

The Apple iPhone 12 is an excellent choice for seniors because it is simple and has numerous functions. It is the best smartphone available for senior citizens willing to spend more than other options. The iPhone 12 Pro Max offers the best performance and design if you don't mind spending more money.

Face ID, dual cameras, wireless charging, and a battery that lasts longer than previous iPhones are among the many features of the iPhone 12. The iPhone SE remains our top pick if you're looking for a less expensive option.

It's easy to see why the iPhone 12 is so popular with seniors. It's stylish, powerful, and loaded with useful features. It is an excellent choice for anyone looking for a dependable smartphone, and Apple makes it very affordable. However, the iPhone 12 is only appropriate for users who aren't concerned with spending much money. If you want something less expensive, the iPhone SE is ideal.

Navigating the iPhone 12 can be difficult for non-technologists. As a result, we've developed a step-by-step guide to assist you in mastering your iPhone 12. Because we've covered all the important information, using the iPhone 12 will be easier after reading this guide.

References

https://www.seniorliving.org/tech/how-to-use-siri/

https://www.phonearena.com/news/how-to-make-iphone-easier-for-seniors-elderly-parents-tutorial_id128365

https://www.macrumors.com/guide/ios-battery-tips/

https://www.apple.com/batteries/maximizing-performance/

https://support.apple.com/en-us/HT207123

https://www.att.com/device-support/article/wireless/KM1273437/Apple/iPhone12Pro

Manufactured by Amazon.ca
Bolton, ON

34518263R00055